MW00947119

My Math Notes for Third Grade

By F. T. Woods

For inquiries, please send an email to: mrwoodsteachesk8@gmail.com

Copyright © 2024 by F. T. Woods, Oceanside, CA.

All rights reserved. No part of this book may be reproduced or transmitted in any form or by any means, electronic or mechanical, including photocopying, recording or by any information storage or retrieval system without written permission or official receipt from the author, except for the inclusion of brief quotations for review.

End of year – Third Grade

Third grade is a pivotal year for math. By third grade students are expected to be able to read to learn and have mastered not only sight and high frequency words, they need to understand academic vocabulary for mathematics. What do third graders need to master in math by the end of third grade? They need to make sense of problems and reason abstractly and quantitively. Students need to model with mathematics, use tools strategically, and attend to precision. Additionally, they need to look for and make use of structure and look for and express regularity in repeated reasoning.

Within this book students will be exposed to math terms, formulas, procedures, vocabulary, and more used within most United States, common core state standards. Additionally, where practical, there will be terms, procedures, vocabulary, and more that can help math be more figure-out-able.

Operations and Algebraic Thinking

- Represent and solve problems involving multiplication and division.
- Understand properties of multiplication and the relationship between multiplication and division.
- Multiply and divide within 100.
- Solve problems involving the four operations, and identify and explain patterns in arithmetic.

Number and Operations in Base Ten

- Use place value understanding and properties of operations to perform multi-digit arithmetic.

Number and Operations—Fractions

- Develop understanding of fractions as numbers.

Measurement and Data

- Solve problems involving measurement and estimation of intervals of time, liquid volumes, and masses of objects.
- Represent and interpret data.
- Geometric measurement: understand concepts of area and relate area to multiplication and to addition.
- Geometric measurement: recognize perimeter as an attribute of plane figures and distinguish between linear and area measures.

Geometry

- Reason with shapes and their attributes.

My 3rd-Grade Notes

Place Value Chart

Converting Words to Numbers and Numbers to Words

Ten-Trillions	Trillions	Hundred-Billions	Ten-Billions	Billions	Hundred Millions	Ten- Millions	Millions	Hundred-Thousands	Ten-Thousands	Thousands	Hundreds	Tens	Ones	And	Tenths	Hundredths	Thousandths	Ten-Thousandths	Hundred-Thousandths	Millionths	Ten-Millionths	Hundred-Millionths	Billionths	Ten-Billionths	HundredBillionths	Trillionths	Ten-Trillionths
5	4,	3	2	7,	6	3	6,	2	7	1,	2	0	5	.	4	9	8	0	8	3	4	7	9	1	4	0	5

fifty-four trillion, three hundred twenty-seven billion, six hundred thirty-six million, two hundred seventy-one thousand, two hundred five and four trillion nine hundred eighty billion eight hundred thirty-four million seven hundred ninety-one thousand four hundred five ten-trillionths

Base-ten Blocks

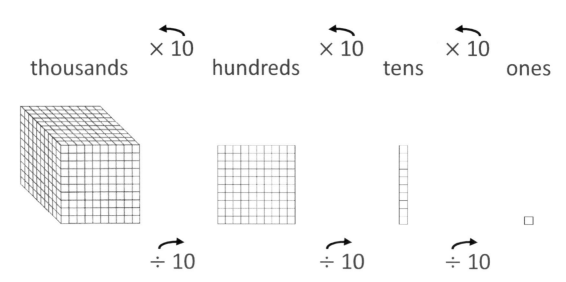

Hundred Chart

1	2	3	4	5	6	7	8	9	10
11	12	13	14	15	16	17	18	19	20
21	22	23	24	25	26	27	28	29	30
32	32	33	34	35	36	37	38	39	40
41	42	43	44	45	46	47	48	49	50
51	52	53	54	55	56	57	58	59	60
61	62	63	64	65	66	67	68	69	70
71	72	73	74	75	76	77	78	79	80
81	82	83	84	85	86	87	88	89	90
91	92	93	94	95	96	97	98	99	100

My 3rd-Grade Notes

Multiplication Table
(Times Table)

X	1	2	3	4	5	6	7	8	9	10	11	12	13	14	15
1	1	2	3	4	5	6	7	8	9	10	11	12	13	14	15
2	2	4	6	8	10	12	14	16	18	20	22	24	26	28	30
3	3	6	9	12	15	18	21	24	27	30	33	36	39	42	45
4	4	8	12	16	20	24	28	32	36	40	44	48	52	56	60
5	5	10	15	20	25	30	35	40	45	50	55	60	65	70	75
6	6	12	18	24	30	36	42	48	54	60	66	72	78	84	90
7	7	14	21	28	35	42	49	56	63	70	77	84	91	98	105
8	8	16	24	32	40	48	56	64	72	80	88	96	104	112	120
9	9	18	27	36	45	54	63	72	81	90	99	108	117	126	135
10	10	20	30	40	50	60	70	80	90	100	110	120	130	140	150
11	11	22	33	44	55	66	77	88	99	110	121	132	143	154	165
12	12	24	36	48	60	72	84	96	108	120	132	144	156	168	180
13	13	26	39	52	65	78	91	104	117	130	143	156	169	182	195
14	14	28	36	56	70	84	98	112	126	140	154	168	182	196	210
15	15	30	45	60	75	90	105	120	135	150	165	180	195	210	225

My 3rd-Grade Notes

Order of Operations

PEMDAS

A mnemonic (ni-mon'ik) device is one that helps with memory. A mnemonic device for remembering the order of operations is **P**lease **E**xcuse **M**y **D**ear **A**unt **S**ally (PEMDAS).

Please
them first.

Parenthesis (or brackets): do everything inside

Excuse

Exponents: do work with exponents.

My **D**ear

Multiplication and Division: do as they occur from left to right.

Aunt **S**ally

Addition and Subtraction: do as they occur from left to right.

NOTE:

PEDMAS is also a way to remember the order of operations.

My 3rd-Grade Notes

Metric System

Relating Units

Each metric prefix is used with meter, gram, and liter to form names for other units of measure.

Prefix	Symbol	Meaning
kilo-	k	1,000
hecto-	h	100
deca-	da	10
deci-	d	0.1
centi-	c	0.01
milli-	m	0.001

1 km = 1,000 m   1 dm = 0.1 m
1 hl = 100 l   1 cg = 0.01 g
1 da = 10 g   1 ml = 0.001 l

To change from a larger unit to a smaller unit, multiply: there will be more smaller units. To change from a smaller unit to a larger unit, divide: there will be less larger units.

Move the decimal point to the right.

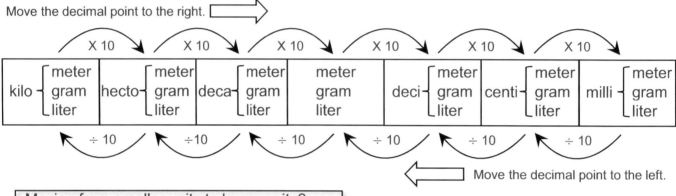

Move the decimal point to the left.

Moving from smaller units to larger units?
Move the decimal point *to the left*.

Example: 6 decimeters = 0.0006 hectometers

Converting from larger units to smaller units?
Move the decimal point *to the right*.

Example: 8.3 kilograms = 8300 grams

Length

1 meter (m) = 100 cm   =   1,000 mm
1 millimeter (mm)   =   0.001m
1 centimeter (cm)   =   0.01 m
1 decimeter (dm)   =   0.1 m
1 decameter (dkm)   =   10 m
1 hectometer (hm)   =   100 m
1 kilometer (km)   =   1,000 m

Capacity

1 liter (l) = 100 cl   =   1,000 ml
1 milliliter (ml)   =   0.001 l
1 centiliter (cl)   =   0.01 l
1 deciliter (dl)   =   0.1 l
1 liter   =   1 l
1 decaliter (dkl)   =   10 l
1 hectoliter (hm)   =   100 l
1 kiloliter (km)   =   1,000 l

Weight

1 gram (g) = 100 cg   =   1,000 mg
1 milligram (mg)   =   0.001 g
1 centigram (cg)   =   0.01 g
1 decigram (dg)   =   0.1 g
1 gram   =   1 g
1 decagram (dkg)   =   10 g
1 hectogram (hg)   =   100 g
1 kilogram (kg)   1,000 g

Standard Measurement

Distance

1 inch (in)	=	2.54 centimeters (cm)
12 in	=	1 foot (ft)
36 in	=	1 yard (yd)
3 ft	=	1 yd
1 meter (m)	=	39.37 in
1 m	=	100 cm
5280 ft	=	1 mile (mi)
1760 yd (yd)	=	1 mile (mi)

Gallon Man

Liquid Measurements (Capacity)

8 ounces (oz)	=	1 cup (c)
2 cups (c)	=	1 pint (pt)
2 pints (pt)	=	1 quart (qt)
4 quarts (qt)	=	1 gallon (gal)

Weight: The amount of force acting on a body (how heavy an object is).

16 ounces (oz)	=	1 pound (lb.)
2000 pounds (lb.)	=	1 ton (t)

Time

60 seconds (sec)	=	1 minute (min)
60 min	=	1 hour (hr)
24 hr	=	1 day (da)
7 da	=	1 week (wk)
4 wk	=	1 month (mo) (approximately)
12 mo	=	1 yr
365 da	=	1 yr (365.25 days, therefore 366 days in a leap year)
10 years (yr.)	=	1 decade (dec)
2 dec	=	1 score
100 yr	=	1 century
1,000 yr	=	1 millennium

Temperatures

32^0 Fahrenheit (F) / 0^0 Celsius (C) =		Freezing Point
212^0 F / 100^0 C	=	Boiling point
98.6^0 F / 37^0 C	=	Average body temperature

My 3rd-Grade Notes

Glossary

addend: Any of the numbers that are added together.

Example: addends $\boxed{5} + \boxed{3} = 8$ sum

algorithm: In math, a step-by-step process to solve a particular problem such as for addition, subtraction, multiplication, and division.

Example: How many groups of 5 are there in 45?

Step 1: Multiply 9 times 5, then write the product under the dividend.

$$5\overline{)\begin{array}{c}9\\45\\45\end{array}}$$

Step 2: Subtract the product from the dividend. Then write the difference below.

$$5\overline{)\begin{array}{c}9\\45\\-45\\\hline 0\end{array}}$$

Step 3: Write the answer as "groups of" the dividend. There are 9 groups of 5 in 45.

a.m. and p.m.: a.m. stands for Ante Meridiem. a.m. is the time from 12-midnight until just before 12-noon. p.m. stands for Post Meridiem. p.m. is the time from noon until just before 12-midnight.

Examples:

1. The time from 12:00 a.m. (Midnight) to 11:59 a.m. is considered the morning.
2. The time from 12:00 p.m. (Noon) to 11:59 p.m. is considered the afternoon/night.

analog clock: A clock (or watch) having moving hands (long: hour hand and short: minute hand) and a circular dial (face), marked with hours and sometimes minutes.

Example: Morning: 10:00 a.m. – Ten o'clock a.m.

Night (afternoon): 10:00 p.m. – Ten o'clock p.m.

angle: The amount of turn, often measured in degrees, between two intersecting lines, around a common point (vertex).

Examples:

acute angle 30^0　　　angle G is obtuse G　　　right angle 90^0

area: The number of unit squares that cover a region, such as a polygon, square, triangle, or circle.

Example: Area (A) of a rectangle is the length (l) times the width (w). A = l × w

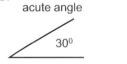 The area of the rectangle is 8m × 2m = 16m²

My 3rd-Grade Notes

array: A way to illustrate (represent, or arrange) numbers or objects, following a specific pattern of columns (up and down) and rows (left to right). An array can be used to show the relationship of factors in multiplication.

Examples:

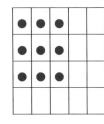

$3 + 3 + 3 = 9$

$3 \times 3 = 9$

♥ ♥ ♥ ♥
♥ ♥ ♥ ♥
♥ ♥ ♥ ♥

$4♥ + 4♥ + 4♥ = 12♥$

$4♥ \times 3♥ = 12♥$

Associative Property of Addition: This property of numbers states that grouping addends in different ways does not change the sum.

Example: $5 + (8 + 7) = (5 + 8) + 7$

Associative Property of Multiplication: This property of numbers states that grouping three or more factors in different ways that the product is the same.

Example: $4 \times (3 \times 2) = (4 \times 3) \times 2 = (4 \times 2) \times 3$

attribute: Any traits or properties of a shape or an object, such as number of sides or angles, lengths of sides, or angle measures. Synonyms: characteristic and feature.

Example:

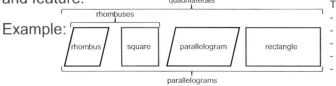

The attributes (properties) of a rhombus:
- Opposite sides are parallel.
- All four sides are equal measure.
- Diagonals bisect each other and are at right angles.
- Opposite angles are equal measure.
- Adjacent angles add up to 180^0.

backward: In reverse or opposite the usual way.

Examples:

1. Count backward to zero starting from 10: 10, 9, 8, 7, 6, 5, 4, 3, 2, 1, 0
2. Counter-clockwise: The opposite direction in which the hands of a clock move.

bar graph: A representation of numerical data in which bars (rectangles) are used to show the number of items in each category by varying the height.

Example: Graph: 2-dogs, 4-cats, and 3-birds

My 3rd-Grade Notes

base-10 blocks: Blocks that are used to represent numbers, usually 1000, 100, 10, and 1.

Example:

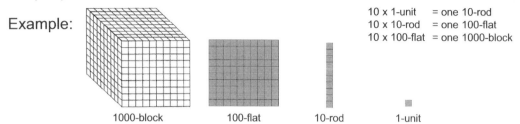

10 x 1-unit = one 10-rod
10 x 10-rod = one 100-flat
10 x 100-flat = one 1000-block

1000-block   100-flat   10-rod   1-unit

beginning (start) time: The exact time when something (action) starts.

Example: Javier went for a run lasting 55 minutes. He finished at 6:15 a.m. When did he start running?

$6:15 - 0:55 = 6:00 - 0:40 = 5:20$ a.m.

Or $6:15 - 1:00 + 0:05 = 5:20$ a.m.

break apart: In math, to separate (decompose) numbers or shapes into parts.

Examples:

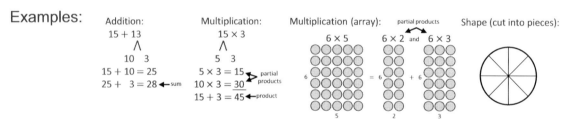

build an array to model multiplication: Draw a picture, use grid paper, or manipulatives. Draw, shade, or place manipulative in columns and rows.

Example: Kara needs to save 30 seats in a movie theater. All the seats need to be together and there can not be more than 6 people in a row. How many rows does she need to save? What I need to find is $6 \times ? = 30$ or $30 \div 6 = ?$

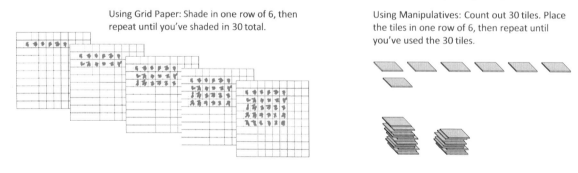

Using Grid Paper: Shade in one row of 6, then repeat until you've shaded in 30 total.

Using Manipulatives: Count out 30 tiles. Place the tiles in one row of 6, then repeat until you've used the 30 tiles.

My 3rd-Grade Notes

calculate: To work out the value of something; To work out an answer using addition, subtraction, multiplication. Note: This does not mean you need a calculator.

Example: Rigo has 15 trees to plant. He's planted five trees. Calculate the difference to find how many more he needs to plant.

The calculation is: $15 - 5 = 10$. The answer is: Rigo needs to plant 10 more trees.

centimeter (cm): A metric unit of measuring length or distance that is equal to one-hundredth of a meter (m). Note, 100cm $=$ 1m and 1cm $=$ 0.01m.

Example:

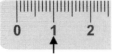

1 centimeter (1cm)

check: To review something, such as math steps, to make sure your work makes sense or is correct.

clockwise: Turn to the right in the direction in which the hands of a clock move.

Example:

classify: To arrange into specific groups, by some property, attribute, or characteristic.

Example: Quadrilaterals have four sides and interior angles add up to 360^0.

quadrilaterals

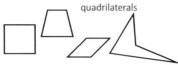

non-example

closed plane: A two-dimensional (2D) figure that is completely enclosed by a boundary and lies completely on a flat surface; A plane shape.

Examples: polygons, circles, and ellipses

Nonexample:

columns and rows: In math, these terms are used to describe the arrangement of numbers or objects in a table, matrix, or array. A column is a vertical (up/down) arrangement of numbers or objects, while a row is a horizontal (left/right or side to side) arrangement of numbers or objects.

Examples:

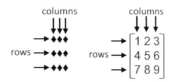

My 3rd-Grade Notes

combined: To put two or more objects or amounts together.

Example: How many different combinations of two animals can you make with dog, cat, and rat?

Answer: (dog, cat), (dog, rat), or (cat, rat)

common denominator: A common multiple of the denominators of two or more fractions.

Example: Some common denominators for $\frac{1}{6}$, $\frac{3}{4}$, and $\frac{1}{2}$ are 12, 24, and 36.

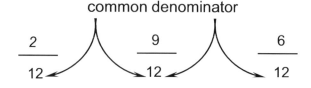

common factor: Factors that are common (same) of two or more numbers.

Example: $\begin{matrix} 6 \\ 3 \times \textcircled{2} \end{matrix}$ ← factors common factor → $\begin{matrix} 8 \\ \textcircled{2} \times 2 \times 2 \end{matrix}$

Nonexample: $\begin{matrix} 4 \\ 2 \times 2 \end{matrix}$ ← factors no common factor (other than 1) → $\begin{matrix} 15 \\ 3 \times 5 \end{matrix}$

common multiple: A number that is a multiple of two or more numbers.

Example: The common multiples of 4 and 6 are 12, 24, 36, and so on.

Commutative Property of Addition: States that the order in which two numbers (addends) are added in different ways does not change the sum.

Example: $(5 + 6) + 7 = 5 + (6 + 7) = (5 + 7) + 6$

Commutative Property of Multiplication: States that the order in which two numbers (factors) are multiplied in different ways does not change the product.

Example: $(5 \times 6) \times 7 = 5 \times (6 \times 7) = (5 \times 7) \times 6$

My 3rd-Grade Notes

compare: A method of comparing two or more numbers, including fractions and decimals, then identifying whether a number is equal, lesser, or greater then the other numbers. Using symbols greater than >; less than <; and equal =.

Examples: $45 > 37$; $2.13 < 2.2$; and $\frac{4}{8} = \frac{1}{2}$

compatible numbers: Numbers that are easy to calculate (compute) mentally. They look nice and friendly.

Example 1: Estimate $29 \div 7$; 28 divides easily

Example 2: Estimate 91×22; 90×20 multiplies easily

compose: Put together numbers to make a greater number.

Example: Compose 3 hundreds, 5 tens, and 7 ones.

Answer: $300 + 50 + 7 = 357$

connect to prior knowledge: Refer to knowledge you've already learned in math, and concepts from previous grades, subjects (reading, history, science, etc.), or daily life.

Example: The costs of video games and gaming consoles.

container: An object that can hold something.

Examples: A cup, hat, pot for a plant, box, bag, etc.

coordinate plane: A two-dimensional surface containing two number lines, one horizontal (right/left) the x-axis and one vertical (up/down) the y-axis.

Example:

My 3rd-Grade Notes

correct: Something that is true or accurate.

Example: 8 is the product of 2 × 4 - correct

Nonexample: 9 is the product of 3 × 4 - incorrect

counterclockwise: Turn to the left in the opposite direction in which the hands of a clock move.

Example:

counting on: A mental math strategy used to add number. Start with the number you're counting on from, then with the other addend, "count on" to get the sum. This can also be used to find the missing addend by counting on then stopping at the known sum.

Example: Find the missing addend: 37 + □ = 45. Start with saying 37, then count on, 38, 39 to 45. Make a hashmark for each number past 37 to and include 45. Add up the hashmarks to find the unknown addend.

cup: Used to measure liquid volume and capacity.

Example: 1 cup (c) = 8 ounces.

data: Facts, numbers, measurements, words, descriptions, quantities, or other things observed.

Example: Vanessa collected rainfall data for a week. What day did it rain nearly three-fourths of an inch?

Answer: On Saturday it rained nearly 3/4 of an inch.

Rainfall: Week 1	
Day	Inches
Monday	0.000
Tuesday	0.012
Wednesday	0.003
Thursday	0.125
Friday	0.340
Saturday	0.748
Sunday	1.001

decide: Think about the situation, problem, action, etc., the possibilities, then choose what you are going to do.

Example:

What is the missing subtrahend in the equation 97 − □ = 34?

I can think about: a. I can count on from 34 up to 97 to determine the subtrahend; or b. I can subtract 34 from 97 to determine the subtrahend. I decided to use strategy b. to find the subtrahend because it's faster.

My 3rd-Grade Notes

decompose (a number): To break apart numbers into two or more parts.

Example: Break apart the number 9. Think of the ways you can break apart 9

- 4 and 5 make 9
- 3 and 6 make 9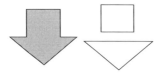
- 1 and 8 make 9

decompose (a shape): To break apart shapes into smaller shapes.

Examples:

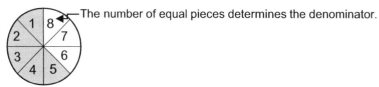

denominator: The number below the bar of in a fraction that indicates the total number of equal parts.

Example: $\dfrac{5}{8}$ denominator

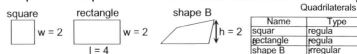

The number of equal pieces determines the denominator.

describe and compare measurable attributes: To see which attributes are in common, see what the differences are (more of; less of; taller; shorter; etc.), then describe, diagram, chart, etc. the differences

Example: Compare the size of and characteristics of quadrilaterals.

square   rectangle   shape B

w = 2   w = 2 / l = 4   h = 2

Quadrilaterals

Name	Type	Heigh
squar	regula	2
rectangle	regula	2
shape B	irregular	2

digit: A lone symbol used to make a numeral (number) used to represent values in math. Digits include 0, 1, 2, 3, 4, 5, 6, 7, 8, 9

Examples

1. What is the digit in the hundreds place value of 235,798? Answer: 7

2. Six is the digit in what place value column of 24.796? Answer: thousandths

digital clock: A digital clock displays the time using digits or other symbols as opposed to an analog clock.

Example:

My 3rd-Grade Notes

Distributive Property of Multiplication: States that multiplying the sum of two or more addends by a number gives the same result as multiplying each addend by the number and then add the products.

Example: $6 \times (9 + 4) = (6 \times 9) + (6 \times 4)$

divide: To arrange a group of things, animals, people, etc. into equal parts.

Example: Divide the blocks into six equal groups.

division: A math operation ($\div$ or $\overline{\smash{\big)}}$) used to separate a number or a number of items (things, people, animals) into equal-size groups.

Example: Amari has a box of chocolates. There are 25 pieces in the box. How many pieces can she share between her and four friends?

Answer: $25 \div 5 = 5$; Amari can give her four friends, five pieces of chocolate each.

dividend: The number being divided by another number.

Examples:

quotient → 7 R1 ← remainder
divisor → 6 ⟌ 43 ← dividend
−42
1

43 ÷ 6 = 7 R1 ← remainder
dividend   divisor   quotient

divisible: Describes if something that can be divided. When dividing one number by another number, if the quotient is a counting number and the remainder is zero.

Example: 27 is divisible by 9 and 3.

divisor: The number that divides the dividend.

Example:

quotient → 7 R1 ← remainder
divisor → 6 ⟌ 43 ← dividend
−42
1

43 ÷ 6 = 7 R1 ← remainder
dividend   divisor   quotient

My 3rd-Grade Notes

dot grid: A type of paper that's similar to grid or lined paper but has evenly spaced dots.

Example: 1 cm dot grid:

elapsed time: The amount of time that has passed (how long) between the start time (beginning) and an end time (end or finish.).

Example: Mr. Wirt ran a 5-kilometer race. He started the race at 6:45 a.m. and finished at 7:05 a.m. What was his elapsed time?

Answer: $7:05 - 6:45 = 20$ minutes. I know because 6:45 to 7:00 is 15 min then add 5 more minutes to get 20 min.

end time: The end time is the start time plus the elapsed time. Also known as the time when an activity finished or stopped.

Example: Janis drove from her home to her job in the town north of her. She started driving at 5:27 p.m. It took her 1-hour and thirty-three minutes to drive. What was the end time that she arrived?

Answer: 5:27 p.m. $+ 1:33 = 7:00$ p.m. Janis arrived at 7:00 p.m.

equation: An algebraic or number sentence which shows that two quantities are equal; A math statement that shows the equality of two expression, by connecting them with an equal sign ($=$).

Example: $5 \times \dfrac{1}{9} = \dfrac{5}{9}$

equivalent expressions: Algebraic expressions that are equal in value; When calculated, the values are the same even though they look different.

Examples:

1. $a(b + 2c) = ab + 2ac$ (Distributive Property of Multiplication)

2. $4x + 9z = 9z + 4x$ (Commutative Property of Addition)

3. $27 \div 3 = 3 \times 3$ (division is the inverse operation of multiplication)

My 3rd-Grade Notes

equivalent fractions: Fractions that represent (name) the same value (amount) after they are simplified (reduced).

Example: $\frac{6}{8}$ and $\frac{3}{4}$ name the same amount.

$\frac{6}{8}$ and $\frac{3}{4}$ are equal amounts.

estimate (verb): Find an answer that is close to the exact amount. **estimate (noun):** A number that is close to an exact amount (tells about how much or about how many).

Examples (verb):

- Estimate a sum: $73 + 28$ is close to $75 + 25 = 100$

- Estimate a difference: $82 - 34$ is close to $82 - 32 = 50$

- Estimate a product: 33×87 is close to $30 \times 90 = 2{,}700$

Example (noun): 6,000 is close to 6,030; 100 pencils is about 97 pencils

even number: An integer that can be divided exactly by 2 (no remainder). The last digit of an even number is always 0, 2, 4, 6, or 8.

Examples: $8 \div 2 = 4$; $957{,}358 \div 2 = 478{,}679$

Non-examples: $9 \div 2 = 4$ R 1; $957{,}359 \div 2 = 478{,}679$ R 1

expanded form: A way to write numbers by showing the value (place value) number for each digit.

Example: $375 = 300 + 70 + 5$

expression: A mathematical sentence (phrase) comprised of numbers (terms, factors, coefficients) and operations signs.

Examples:

- $(60 + 40) \div 50$

- $5 - 7y$

My 3rd-Grade Notes

fact family: A collection of math facts that show the relation between the same set of numbers. There are two types of fact families. Also known as a number bond or fact family triangle.

Examples:

Addition and Subtraction Fact Family:

 The numbers 2, 4, and 6 form a fact family.
- $2 + 4 = 6$
- $4 + 2 = 6$
- $6 - 4 = 2$
- $6 - 2 = 4$

Multiplication and Division Fact Family:

 The numbers 2, 4, and 8 form a fact family.
- $2 \times 4 = 8$
- $4 \times 2 = 8$
- $8 \div 4 = 2$
- $8 \div 2 = 4$

factor: A number that is multiplied by another number to find a product.

Example: $3 \times 8 = 24$

     factor   factor

figure: Two- (2D) and three-dimensional (3D) shapes.

Examples: 2D figure - square   3D figure - cube

flat shape (plane shape): A two dimensional (2D) shape having only two dimensions i.e. length and width or width and height; A shape that can be drawn on any flat surface or plane paper.

Examples: circle, diamond, hexagon, kite, pentagon, rectangle, and triangle

My 3rd-Grade Notes

fluid ounce (fl. oz.): A US measurement to measure liquid volume and capacity. 1 fl. oz. is nearly equal to 29.57 milliliters (mL).

Example: 8 fl. oz. = 1 cup (c)

formula: A group of math symbols that express a mathematical rule, relationship, or that are used to solve a problem.

Example: Area (of a rectangle) = base × height, or A = b × h

fourth: The same as the fraction $\frac{1}{4}$ and decimal 0.25; one-quarter of an amount when divided by four.

Examples: $\frac{1}{4}$ is not shaded.

denominator

Three-fourths (3/4) of the circle are shaded.

fraction: A number that represents (names) a part of a whole or part of a group.

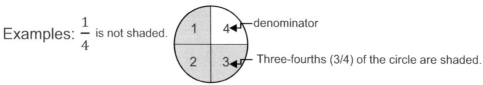

fraction bar

Example: $= \frac{1}{4} =$

gallon (gal): One US gallon (gal) is a unit for measuring liquid volume and capacity. Additionally, 1 US gal is 128 fl. oz. and equals 231 cubic inches (231 in^3) or about 3.785 liters.

Example: 1 gal = 8 pints = 4 quarts = 16 cups = 128 fl. oz.

gram (g): A metric unit of measurement of mass (weight).

Examples: A paperclip weighs about one gram (1g). A kilogram is 1,000g.

My 3rd-Grade Notes

greater than (>): Used to compare the relationship between two numbers, decimals, fractions, quantities, mass, etc. The symbol > is used to compare greater on the left and less on the right.

Examples: 256 > 128; 1.1 > 1.02; $\frac{1}{2} > \frac{2}{8}$; 5-bananas > 3-oranges, 10g > 9.9g

half hour: One half of one hour (60 min) is 30 minutes (min).

Example: Half past 3 p.m. is a half hour past 3 p.m.

half gallon (half-gal or ½ gal): One-half US gal is a unit for measuring liquid volume and capacity. Additionally, 1/2 US gal is 64 fl. oz. or about 1.893 liters.

Example: 1 half gal = 4 pints = 2 quarts = 8 cups = 64 fl. oz.

halfway: At the middle of (between) two points; midway.

Example: Circle then name the fraction that is halfway between zero and one?

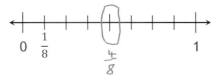

halves: One of two equal parts of a quantity.

Example: Draw a line to divide the shapes into halves.

hexagon: A closed two-dimensional (2D) polygon with six sides.

Examples:

regular hexagon　　　irregular hexagon

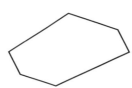

My 3rd-Grade Notes

hour (h): A unit of time that measures 60 minutes; there are 60 min in 1 h.

Example: How long has passed between Clock A and Clock B?

Clock A 　　Clock B 　Answer: One hour has past.

hour hand: The short (small) hand on an analog clock.

Examples:

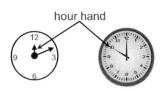

improper fraction: A fraction with a numerator (top number) that's larger than or equal to the denominator (bottom number); The fraction is "top heavy."

Example: $\frac{9}{7}$ nine-sevenths is a "top heavy" fraction.

incorrect: Something that is neither true nor accurate.

Example: 9 is the product of 2×4 - incorrect

Nonexample: 12 is the product of 3×4 - correct

integer: A number with no fractional part; a positive or negative counting number.

Examples: 9, 54, -73, 268, 0, -91

Nonexamples: 1.003, -3.25, $\frac{3}{4}$

inverse operations: Operations that undo each other (opposite operations), such as addition and subtraction or multiplication and division.

Example: $7 \times 8 = 56$ and $56 \div 8 = 7$; $9 + 6 = 15$ and $15 - 9 = 6$

My 3rd-Grade Notes

key: A section that defines or explains the symbols used in a graph.

Example: More raffle tickets were sold on Thursday.

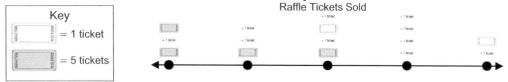

label (verb): To assign an attribute (classify) such as a word, word phrase, letter, number, or a combination of the previous. label (noun): attach a small piece of paper, cloth, sticker, or similar material, with information on it, to an object (example: price tag).

Example (verb): The length is three times the width of the rectangle. Label the width.

$w = 3$

$l = 9$

length: The size of something (an object) or distance from one point to another.

Example: Answer: The length of the rectangular prism is three meters.

$l = 3$ m

less than (<): Used to compare the relationship between two numbers, decimals, fractions, mass, etc. The symbol < is used to compare less on the left and greater on the right.

Examples: $128 < 256$; $1.02 < 1.2$; $\frac{2}{16} < \frac{2}{8}$; 5-bananas < 9-oranges, 8.9g < 9g

letter for an unknown number: A letter can be used for an unknown value in a math sentence. Also known as a variable.

Example: If the area of a rectangle is 125 units, and the length is 25 units, write a multiplication and division equation that shows the unknown value as B.

Answer: $A = l \times w$, so $125 = 25 \times B$; $B = 125 \div 25$

line plot: Is a graph that displays (records) data using symbols representing a piece of data on a number line.

Example:

liquid volume: The measure of the space a liquid is contained in.

Examples:

US units of liquid volume: 1 gal = 8 pints = 4 quarts = 16 cups = 128 fl. oz.

Metric units of liquid volume: 1 liter (L) = 1,000 milliliters (mL) ≈ 33.81 fl. oz.

liter (L): A metric unit of volume used to measure liquids.

Example: 1 liter (L) = 1,000 milliliters (mL) ≈ 33.81 fl. oz.

mass: A measure of how much matter is in an object.

Example: Body mass data for third graders show the average weight is 69.8 pounds or about 31.7 kilograms.

measure: Use know standards such as a ruler, measuring cup, square unit, or scale, to find the length, liquid volume, area, or mass of an object.

Example: Find the length of the tube for the dropper from 0.25ml to 1ml.

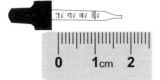

Answer: the measure of the dropper
from 0.25ml to1ml is 1cm.

measure area using unit squares: Use unit squares to measure the area of a rectangle or rectangles are that are added.

Example: Find the area of the shape below using square units (sq. units).

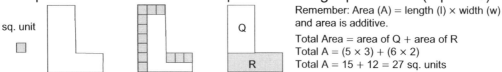

Remember: Area (A) = length (l) × width (w) and area is additive.

Total Area = area of Q + area of R
Total A = (5 × 3) + (6 × 2)
Total A = 15 + 12 = 27 sq. units

meter (m): A unit for measuring length or distance. One meter equals 100 centimeters.

Example:

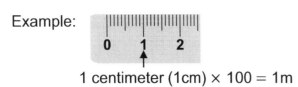

1 centimeter (1cm) × 100 = 1m

My 3rd-Grade Notes

midnight: The middle of the night when one day ends and a new day starts; both the hour hand and minute hand points directly to 12; 12 midnight.

Example:

mile (mi): A unit for measuring length or distance. One mile equals 5,280 feet (ft).

Example: 1 mi = 5,280 ft = 1.609.34 meters (m).

milligram (m): A small metric unit for measuring weight/mass; one-thousandths of a gram ($0.001g = \frac{1}{1000}g$).

Example: A paperclip weighs about 1g or 1000mg.

milliliter (mL): A small metric unit that represents volume of the capacity of a liquid.

Example: A graduated medicine dropper is usually measured in milliliters.

millimeter (mm): A metric unit of measurement used to measure the length of small or tiny objects.

Example:

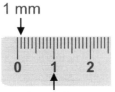

10 mm = 1 centimeter (1cm)

min (min): A measurement of time that equals sixty-seconds.

Example: There are 60 min in one hour.

My 3rd-Grade Notes

minute hand: The long (big) hand on an analog clock.

Examples:

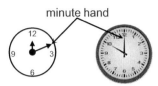

minutes after and minutes before: The number of minutes after (past) the hour or the number of minutes before (until) the next hour.

Examples: 8 min before 10   6 min after 12

mixed number: Consists of a whole number and a proper fraction.

Example: $2\frac{5}{16}$ is a mixed number

whole number part   fraction part

$$2\frac{5}{16}$$

model: An object or picture used or created to represent a situation or real world things.

Example: A pack of colored pencils contains 7 pencils. Each pack costs $3. If you need six of each color, how many pencils in all will you have? Create a model to show your thinking.

$6 \times 7 = 42$

multiple: The product of a whole number and another whole number is called a multiple of that whole number.

Example:

10	10	10	10	whole numbers
× 5	× 4	× 3	× 2	
50	40	30	20	multiples of 10 (and by extension 5 and 2)

multiply (multiplication): To repeatedly add the same number a certain number of times; find the total number (product) of things in equal groups.

Example: Aaron has a book of stamps. On each page he has nine stamps. So far he's collected enough stamps to fill 8 pages. How many stamps does Aaron have? $9 \times 8 = 72$, Aaron has 72 stamps in all.

multi-step word problem: A word problem that has more than one step to complete.

Example: Francesca has thirty-six bows. Six bows are white, two are green, and the rest are pink. Melannie has twenty-nine bows. Four are blue, one is red, and the remainder are pink. Together, how many pink bows do Francesca and Melannie have?

Step 1: Find how many pink bows Francesca has. 36 bows − 6 white bows − 2 green bows = 28 pink bows.

Step 2: Find how many pink bows Melannie has. 29 bows − 4 blue bows − 1 red bow = 24 pink bows.

Step 3: Add Francesca's bows and Melannie's bows. 28 + 24 = 52 pink bows. Franchesca and Melannie together have 52 pink bows.

n square unit: Any square unit used to measure area.

Example of square measures:

- square inches (in^2)
- square feet (sq ft or ft^2)
- square meters (m^2)
- square centimeters (cm^2)

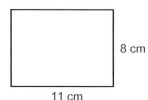

8 cm

11 cm

The area of the rectangle is:
11 cm × 8 cm = 88 cm^2

non-rectangular shapes: Shapes that do not have four straight sides and four right angles.

Examples:

rectangular shapes

non-rectangular shapes

noon: The middle of the day (midday); both the hour hand and minute hand points directly to 12; 12 noon.

Example:

number line: A straight line drawn with numbers placed, in order, and equal distances along it's length.

Examples:

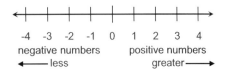

-4 -3 -2 -1 0 1 2 3 4
negative numbers   positive numbers
◄——— less   greater ———►

Skip count to determine how many eighths are where the arrow is pointing.

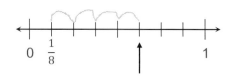

0    $\frac{1}{8}$ 1

My 3rd-Grade Notes

numerator: The number above the bar of in a fraction that indicates the number of selected parts.

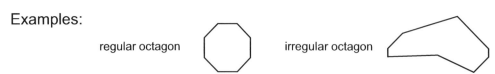

Example: $\dfrac{5}{8}$ ← numerator   The number of selected (shaded) parts

octagon: A polygon (flat shape) with eight sides and eight angles; a regular octagon has eight equal sides and eight equal angles.

Examples:

regular octagon   irregular octagon

odd number: Numbers that cannot be divided into two equal groups; numbers that are not divisible by 2 (have a remainder); The last digit of an odd number is never 0, 2, 4, 6, or 8.

Examples: $9 \div 2 = 4$ R 1; $957{,}359 \div 2 = 478{,}679$ R 1

Non-examples: $8 \div 2 = 4$; $957{,}358 \div 2 = 478{,}679$

one-fourth: A proper fraction where the numerator is one and the denominator is four; one-quarter; a fourth; 1/4.

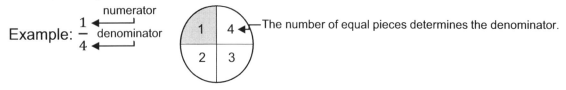

Example: $\dfrac{1}{4}$ ← numerator / denominator   The number of equal pieces determines the denominator.

one-half: A proper fraction where the numerator is one and the denominator is two; a half; 1/2.

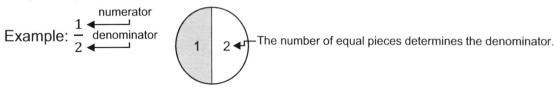

Example: $\dfrac{1}{2}$ ← numerator / denominator   The number of equal pieces determines the denominator.

one-third: A proper fraction where the numerator is one and the denominator is three; a third; 1/3.

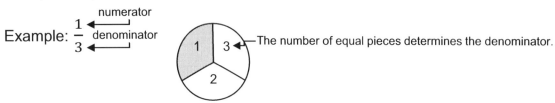

Example: $\dfrac{1}{3}$ ← numerator / denominator   The number of equal pieces determines the denominator.

My 3rd-Grade Notes

one-step word problem: A word problem that has only one step to complete.

Example: Jason has 15 model cars. Jesus has 12 model cars. How many more model cars does Jason have than Jesus?

Step 1: Find the difference of cars between Jason and Jesus: $15 - 12 = 3$. Jason has 3 more cars than Jesus.

ordering (sorting): Arranging numbers, objects, and things in a defined sequence such as increasing or decreasing value.

Example: Order the following numbers in decreasing value. 391, 847, 621, 109

Answer: 847, 621, 391, 109

organizing quantities: Make a table or list to organize data in a problem. This is useful to find relationships and patterns among data.

Examples:

How many ways can you arrange the letters W, X, and Y?
WXY, WYX, XWY, XYW, YWX, YXW

Gio wants to have exactly 16 cents. He only has pennies and nickels. What are all the ways to do that?

Number of Nickels	Number of pennies
0	16
1	11
2	6
3	1

ounce (oz.): A unit measure of mass (weight). Note: this is different from fl. oz. that measures volume.

Example: A standard pencil is about an ounce.

parallel: Always the same distance apart.

Example: Parallel lines are in the same plane, always the same distance apart, and never intersect (meet).

parallelogram: A parallelogram is a two-dimensional geometrical shape, whose opposite sides are parallel to each other. It is a type of polygon having four sides (also called quadrilateral), where the pair of parallel sides are equal in length.

Example:

Note: A rhombus is a parallelogram but a parallelogram is not a rhombus.

paraphrase: (verb) Say something another way; (noun) a rewording of something written or spoken.

Example: Original problem: "There are 12 apples in a basket. If you take away 5 apples, how many apples are left?"

Paraphrased problem: "You have 12 apples. Five apples are removed, how many apples are left?"

partial product: A method for solving multiplication problems where ones, tens, hundreds, and so on are multiplied separately and then the separate products are added together to find the product.

Example:
$$\begin{array}{r} 14 \\ \times\quad 5 \\ \hline 50 \\ 20 \\ \hline 70 \end{array}$$

50, 20 are the partial products

partial quotient: A method for solving division problems using repeated subtraction (subtract multiples of the divisor from the dividend) to find the quotient.

Example:

partial quotients

$$\begin{array}{r} 5\overline{)155} \\ -\ 50 \\ \hline 105 \\ -\ 50 \\ \hline 55 \\ -\ 50 \\ \hline 5 \\ -\ 5 \\ \hline 0 \end{array}$$

10×5 10
10×5 10
10×5 10
1×5 +1
      31

partial sum: A method for solving addition problems where ones, tens, hundreds, and so on are added separately and then the separate sums are added together to find the sum.

Example:
$$\begin{array}{r} 43 \\ +\quad 22 \\ \hline 60 \\ 5 \\ \hline 65 \end{array}$$

$40 + 20 = 60$
$3 + 2 = 5$
60, 5 are the partial sums

partition: Split numbers into smaller parts to make them easier to work with; break into parts; sometimes break into equal parts.

Examples: Partition the number 63 into 2 parts and 4 parts

2 parts: $63 = 60$ and 3

4 parts: $63 = 20$ and 20 and 20 and 3

pattern: An ordered set (or sequence) of numbers, shapes, colors, or other objects that follow a rule.

Examples: 3, 6, 9,12 or ♦♥♦♥♦♥

2 times table: 2, 4, 6, 8 … each number is increased by 2

5 times table: 5, 10, 15, 20 … each number is increased by 5

My 3rd-Grade Notes

perimeter (P): The total distance of the boundary (around) a shape.

Example: P = 4 cm + 3 cm + 4 cm + 3 cm = 14 cm

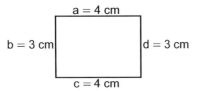

Note: Perimeter formula for rectangles, squares, and parallelograms

P = 2a + 2b

pentagon: A five-sided polygon on a plane. A regular pentagon is a polygon with five sides of equal length and five angles of equal measure.

Example:

regular pentagon   irregular pentagon

picture graph (pictograph): An illustration of data using pictures or symbols to represent quantities.

Example:

Number of Bananas Eaten by grade level in a day

Grade Level	Number of Bananas
1	🍌🍌🍌
2	🍌🍌
3	🍌🍌🍌🍌
4	🍌

🍌 = 10 Bananas

plane figure: A two-dimensional (2D) shape that can be drawn on a flat surface; a figure that has length and width and no thickness (height). See **closed plane**, page 4.

Examples: Includes rectangles, triangles, polygons, circles, and ellipses …

plot ordered pairs: An ordered set (or sequence) of numbers, shapes, colors, or other objects that follow a rule.

Examples: Plot the ordered pairs (1,2) and (3,4), then draw a line between them.

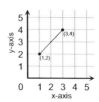

polygon: A two dimensional (2D)shape, that is made of straight lines, and is closed. Note: See **closed plane** and **closed figure**.

Examples:

All are closed and have straight lines.

Nonexamples:

    Have curved lines.

    Open and not closed.

My 3rd-Grade Notes

product: The result (answer) of multiplying two or more numbers together (multiplication sentence).

Example: $6 \times 7 = \boxed{42}$ ◄——— product

quadrilateral: A polygon with four sides and four angles.

Examples:

quart (qt): A measure of unit capacity or liquid volume.

Example: 1qt = 2pt = 4c = 32 fl. oz.

quarter after (past): A measure of time, 15 minutes past the hour.

Examples:

quarter before (until): A measure of time, 15 minutes before the hour.

Examples: quarter before 10   quarter before 1

quarter hour: A measure of time of 15 minutes.

Example: Carmelo left his house to walk to school at 8:20 a.m. He arrived at school at 8:35 a.m. It took him a quarter hour to walk from his home to school.

quotient: The result (answer) when one number is divided by another number, not including the remainder.

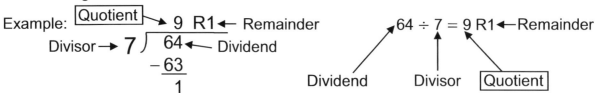

Example:

| Quotient → 9 R1 ← Remainder |
| Divisor → 7) 64 ← Dividend |
| − 63 |
| 1 |

64 ÷ 7 = 9 R1 ← Remainder

Dividend   Divisor   Quotient

real world math problem: A question (math problem) relating to a concrete (real) setting.

Example: A farmer has 13 cows. Each cow gives 4 gallons of milk each day. How many gallons of milk do all the cows give in 3 days?

record: To capture or write something (information or data) down so that it can be used again.

Example: To find the total rainfall for a week, you need to record the amount it rained every day.

regroup: Rearranging numbers into groups by place value; or exchange amounts of equal value to rename a number; or carrying (in addition) and borrowing (in subtraction).

Example: 6 + 9 = 15 ones or 1 ten and 5 ones.

six ones plus nine ones equals fifteen ones

relate area and perimeter: Area is the pace occupied by a shape. Perimeter is the distance of the boundary of the shape. As the area of the shape increases, the perimeter of the shape increases.

Example:

a = 4 cm

b = 3 cm

$P = 2a + 2b = 2(4 \text{ cm}) + 2(3 \text{ cm}) = 14 \text{ cm}$

$A = a \times b = 4 \text{ cm} \times 3 \text{ cm}) = 12 \text{ cm}^2$

Therefore: If any side of the rectangle increases or decreases the area will change.

relationship between even and odd numbers: Even and odd numbers are whole numbers. If you multiply an even number by an odd or even number you will get an even number. If you add two odd numbers or two even numbers together you will get an even number. If you add an even number with an odd number you will get an odd number. Can you figure out the rules for the inverse?

Examples: $4 \times 3 = 12$; $4 \times 6 = 24$; $9 + 9 = 18$; $22 + 32 = 54$; $8 + 9 = 17$

My 3rd-Grade Notes

remainder: The amount left over (after performing division) when a number cannot be divided equally.

Example:
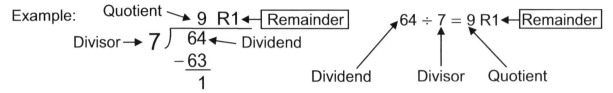

repeated addition: The process of adding the same number over and over again. Understanding what repeated addition is will help with understanding and figuring out multiplication.

Example: $5 + 5 + 5 + 5 + 5 = 25$; where 5 added five times is the same as, 5 times 5: $5 \times 5 = 25$

represent: To use something such as a sign, symbol or drawing to stand for something else.

Examples:

• Five times five can be represented as, 5×5

• A missing factor (product, addend, etc.) can be represented as a letter, $5 \times b = 20$

rhombus: A flat shape with four equal length and straight sides. Opposite sides are parallel and opposite angles are equal. It is a parallelogram and a quadrilateral. A square is also a rhombus but with all right angles.

Example:

right angle: An angle whose measure is 90^0 (forms a square corner).

Example:

90^0

round: Replacing a number with a simpler number or a number that is close to how many or how much.

Examples:

Round 454 to the nearest ten = 450

Round 37.358 to the nearest hundredth = 37.360 or 37.36

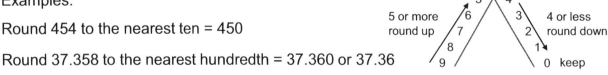

Remember: If the number to the right is 5 or more, raise the score. If four or less, let it rest.

My 3rd-Grade Notes

rows and columns: In math, these terms are used to describe the arrangement of numbers or objects in a table, matrix, or array. A column is a vertical (up/down) arrangement of numbers or objects, while a row is a horizontal (left/right or side to side) arrangement of numbers or objects.

Examples:

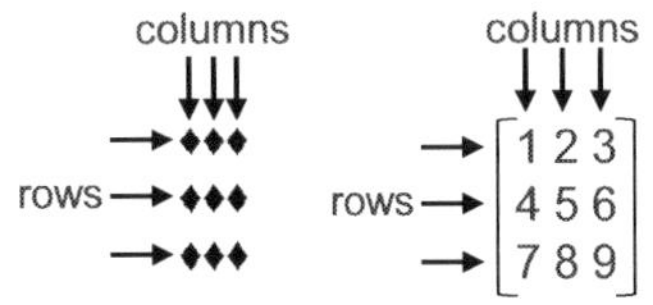
columns
rows
columns
rows
1 2 3
4 5 6
7 8 9

rule: An explicitly specified way one follows to go from one number or shape to the next in a pattern.

Example: Rule – add 3 then multiply the sum by 2 given the number 2:

The next number is $(2 + 3) \times 2 = 10$; next is 26, 58, 122, ...

scaled bar graph: A graph that represents data, quantities, values, or numbers using bars or strips.

Example: Compare and contrast how many 3rd graders like which seasons of the year.

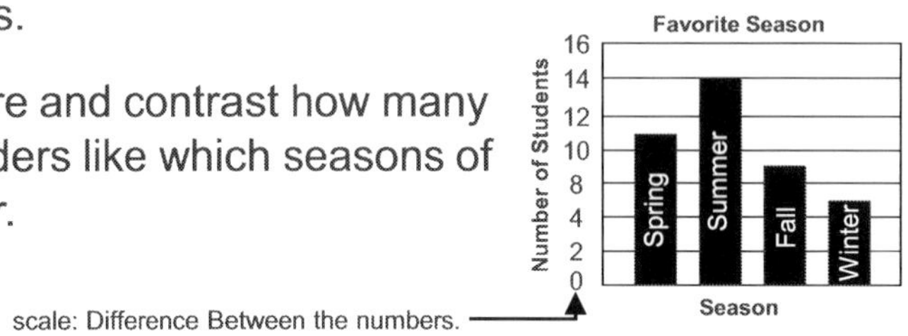
Favorite Season
Number of Students
16
14
12
10
8
6
4
2
0
Spring
Summer
Fall
Winter
Season
scale: Difference Between the numbers.

second (sec): A small unit of time.

Examples: 1 minute = 60 sec; 1 sec = 1/60 min

sequence: A list of objects or numbers following a specific pattern or rule.

Example: Start a sequence where the first term is 3. Add one to the term, then add the term to that sum to find the next term. Stop at the fourth term.

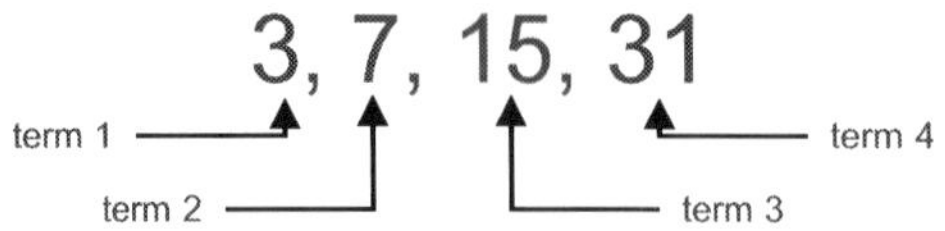
3, 7, 15, 31
term 1
term 2
term 3
term 4

solution: The answer to a problem; A value, or values, that can replace a variable (unknown) to make an equation true.

Examples:

A. What is the missing symbol in the sequence
☻☼☺_☻☼☺◘

The solution is: ◘

B. Find y where $y + 7 = 26$

$$y + 7 = 26$$
$$-7 = -7$$

The solution is: $y = 19$

sort: To separate, arrange, or group objects (things, numbers, shapes, etc.) in a certain way such as by size, color, type, numerically, alphabetically, etc.

Example: Sort the rectangles by width, from least to greatest.

square unit: The product of area of a square with 1 unit per side; or a unit of area with dimensions of 1 unit × 1 unit.

Example:

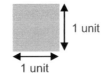

standard form: A way of representing (writing) numbers by using the digits 0-9, using place value for each digit.

Example: Two thousand, three hundred, forty-five is written in the standard form as 2,345.

sum (total): The solution when adding two or more numbers.

Example: Jayne has three cats and Byrum has four cats. How many cats do they have together?

$$3 + 4 = 7$$

addend — addition operator — addend — sum

Answer: Jayne and Byrum have 7 cats together.

term: 1. Values on which math operations appears in an algebraic expression; 2. A number or object in a pattern.

Examples: 1. 6 +7y

term   term

2. ♦♥♦♥♦♥ or 3, 6, 9, 12

terms   terms

three factor expression: An expression with three factors.

Examples: 5 × 3 × 6 is an expression that has three factors.

Using the Associative
Property of Multiplication:

$(5 × 3) × 6 = 5 × (3 × 6)$

Using the Commutative
Property of Multiplication:

$(5 × 3) × 6 = (5 × 6) × 3$

My 3rd-Grade Notes

ton (T): A measure of unit weight (mass). Note: This is different from a metric ton.

Examples: 1T = 2,000 Lbs. A small car is about 1 ton.

trapezoid: A quadrilateral with one pair of parallel sides (exclusive); A parallelogram (quadrilateral) with two pairs of parallel sides (inclusive).

Examples:

trapezoid
(exclusive)

trapezoid
(exclusive, quadrilateral)

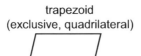

triangle: A closed, flat shape (2D), with three straight sides, and three interior angles.

right triangle   equilateral triangle   isosceles triangle   scalene triangle

Examples:

two-step word problem: A word problem that requires two steps to find the answer (solve). You can solve a two step problem with two one-step equations or one two-step equation.

Example: Huey has 48 dog treats. He gives 3 treats to Charlie the dog. He will then give the same amount to other dogs. How many other dogs can Huey give treats to?

Step 1: $48 - 3 = 45$ Step 2: $45 \div 3 = 15$ Answer: Huey can give 3 treats to 15 other dogs.

unit fraction: A fraction with a numerator of 1 and the denominator is any natural number.

Example:

unit fraction

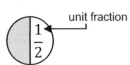

unknown addend: An addition problem with one known addend and the sum.

Examples: Find the missing (unknown) addends.

$$32 + ? = 45 \rightarrow ? = 45 - 32 = 13$$

$$X + 9 = 37 \rightarrow X = 37 - 9 = 28$$

$$91 + \square = 43 \rightarrow \square = 91 - 43 = 48$$

Notice the different symbols used for the missing addend.

My 3rd-Grade Notes

unknown factor: A multiplication problem with one known factor and the product.

Examples: Find the missing (unknown) addends.

$$32 \times \ ? = 96 \rightarrow \ ? = 96 \div 32 = 3$$

$$Y \times 9 = 81 \rightarrow Y = 81 \div 9 = 9$$

Notice the different symbols used for the missing factor.

use an area model to show fractions: An area model is drawn to represent a fraction or fractions.

Examples:

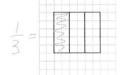

volume: The measure of space inside a three-dimensional (3D) figure.

Example: Volume (V) of a rectangular prism is length times width times height or $(l \times w \times h)$.

$$4 \text{ cm} \times 4 \text{ cm} \times 3 \text{ cm} = 48 \text{ cm}^3$$

w = 4 cm

h = 3 cm

l = 4 cm

whole number: A set of positive numbers starting at zero that increases by one for each digit; neither a fraction, decimal, nor negative number.

Example: A set of whole numbers include 0, 1, 2, 3, 4, 5 , 6, 7, 8, 9, 10, 11, …

whole number as a fraction: The fraction represents a whole.

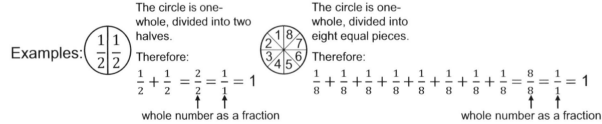

Examples:

The circle is one-whole, divided into two halves. Therefore:

$$\frac{1}{2} + \frac{1}{2} = \frac{2}{2} = \frac{1}{1} = 1$$

↑ ↑
whole number as a fraction

The circle is one-whole, divided into eight equal pieces. Therefore:

$$\frac{1}{8} + \frac{1}{8} + \frac{1}{8} + \frac{1}{8} + \frac{1}{8} + \frac{1}{8} + \frac{1}{8} + \frac{1}{8} = \frac{8}{8} = \frac{1}{1} = 1$$

↑ ↑
whole number as a fraction

Zero Property of Multiplication (Zero Property): When any number is multiplied by zero, the product is always zero.

Examples:

$$199 \times 0 = 0$$

$$Y \times 0 = 0$$

Note: Any number, fraction, decimal, or unknown number multiplied by zero is zero.

My 3rd-Grade Notes

whole: All of the parts of a share or group.

Example: All of the parts of a 100 grid is 100.

100 grid

word form: Express (write) numbers using words.

Example: 342,125 is written as three hundred forty-two thousand, one hundred twenty-five

My 3rd-Grade Notes

My 3rd-Grade Notes

Made in the USA
Las Vegas, NV
01 February 2024

85177464R00046